Animals of Africa

Giraffe

Katie Gillespie

AV2
www.av2books.com

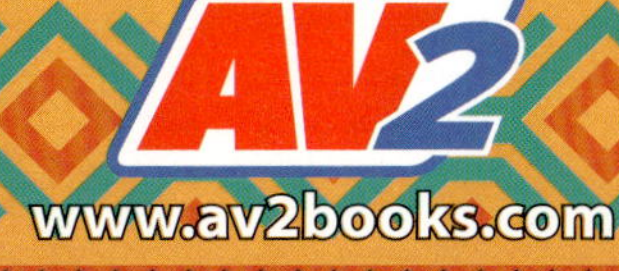

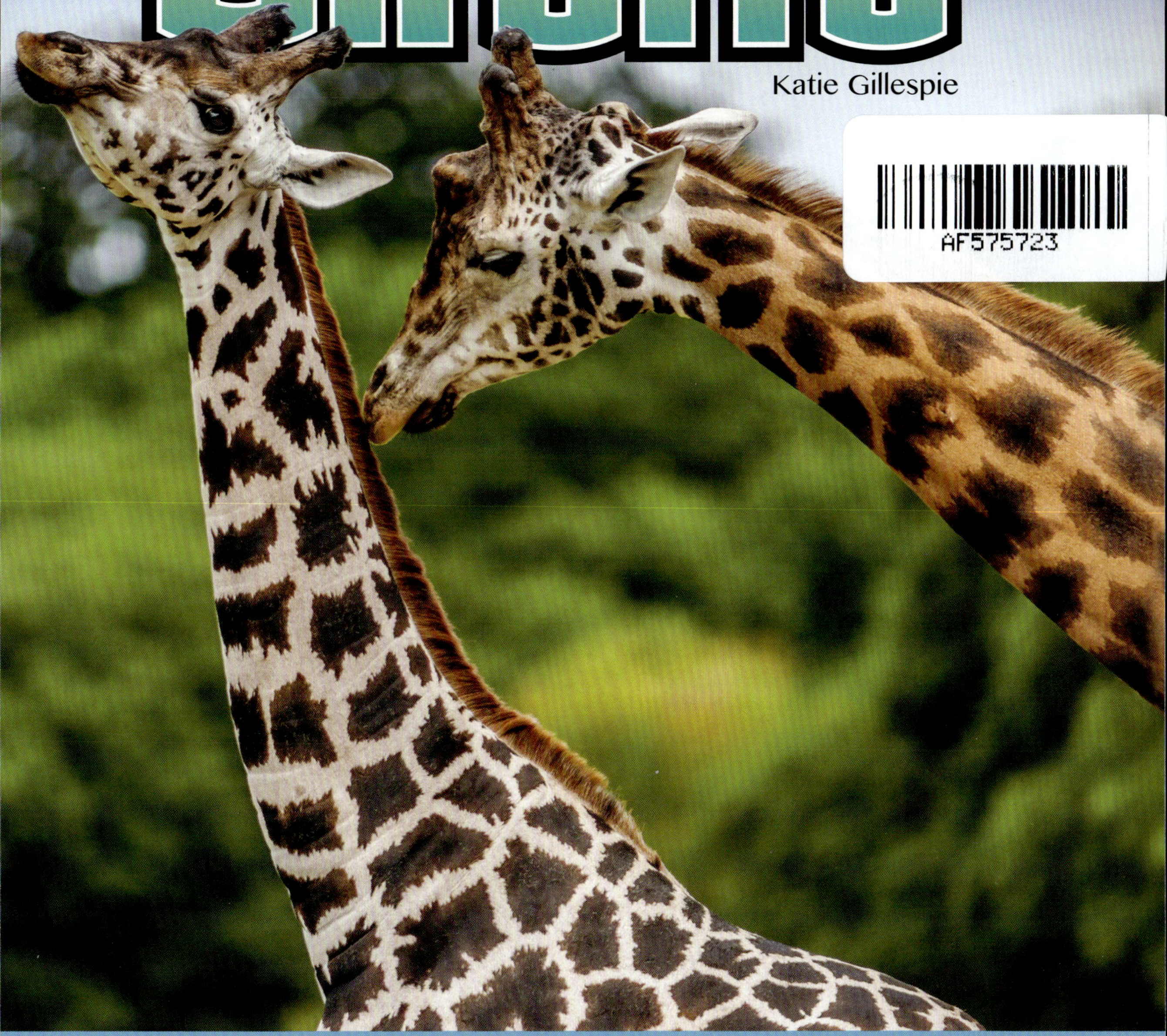

Step 1
Go to **www.av2books.com**

Step 2
Enter this unique code
FBYLD2700

Step 3
Explore your interactive eBook!

AV2 is optimized for use on any device

Your interactive eBook comes with...

Contents
Browse a live contents page to easily navigate through resources

Audio
Listen to sections of the book read aloud

Videos
Watch informative video clips

Weblinks
Gain additional information for research

Try This!
Complete activities and hands-on experiments

Key Words
Study vocabulary, and complete a matching word activity

Quizzes
Test your knowledge

Slideshows
View images and captions

This title is part of our AV2 digital subscription

1-Year Grades K–5 Subscription
ISBN 978-1-7911-3320-7

Access hundreds of AV2 titles with our digital subscription.
Sign up for a FREE trial at **www.av2books.com/trial**

Giraffe

CONTENTS

Meet the Giraffe

Most giraffes live in central and southern Africa. They make their homes on the open plains. These areas provide them with food to eat and room to roam.

There is only one **species** of giraffe, but there are nine different **subspecies**. Although the different kinds of giraffes vary in height and coloring, all share common features. Every giraffe has a long neck, big eyes, hoof-like feet, and two horn-like growths on its head.

A Tall Animal

Giraffes are the tallest animals on land. They can grow to be 18 feet (5.5 meters) tall and weigh as much as 2,800 pounds (1,270 kilograms). A giraffe's neck makes up almost half of its height.

Giraffes are so tall that they could look directly into a second-story window. Their great height allows giraffes to eat food that other animals cannot reach. It also helps them look out for **predators**.

Comparing Heights

GIRAFFE
18 feet
(5.5 meters)

DOUBLE-DECKER BUS
14 feet
(4.3 meters)

Keeping Together

Females and young giraffes usually stay together in groups called herds. Each herd is made up of about six giraffes. Male giraffes are more **solitary**.

Giraffes move around constantly in search of food. They can run up to 35 miles (56 kilometers) per hour for short periods of time. They can run for longer stretches at 10 miles (16 km) per hour.

Growing Up

A giraffe gives birth while standing up. Its baby is called a calf. A female giraffe carries its calf for 15 months

Calves look just like adult giraffes, only smaller. They stand about 6 feet (2 meters) tall at birth. A calf drinks milk until it is 4 months old. It then begins to eat leaves. Young giraffes stay with their mothers until they are 2 years old.

BIG FACT
A giraffe calf can run just 10 hours after it is born.
10 hours

A Closer Look

Giraffes have **adapted** to their **habitat**. They have unique features to help them survive in their environment.

Neck

The giraffe's long neck is used to reach food found high in treetops.

Patterned Coat

The pattern on a giraffe's coat lets the giraffe blend into its environment.

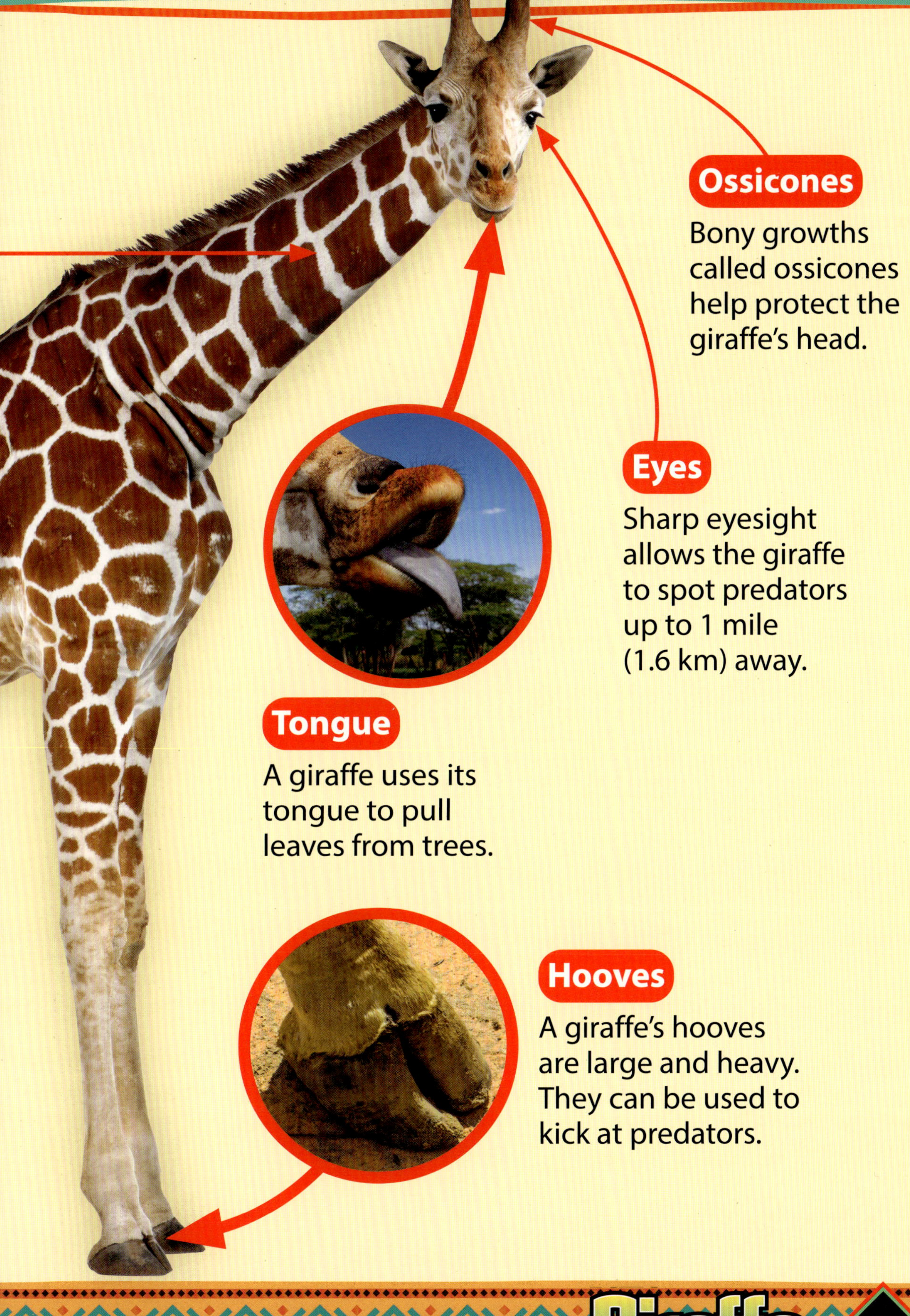

Ossicones

Bony growths called ossicones help protect the giraffe's head.

Eyes

Sharp eyesight allows the giraffe to spot predators up to 1 mile (1.6 km) away.

Tongue

A giraffe uses its tongue to pull leaves from trees.

Hooves

A giraffe's hooves are large and heavy. They can be used to kick at predators.

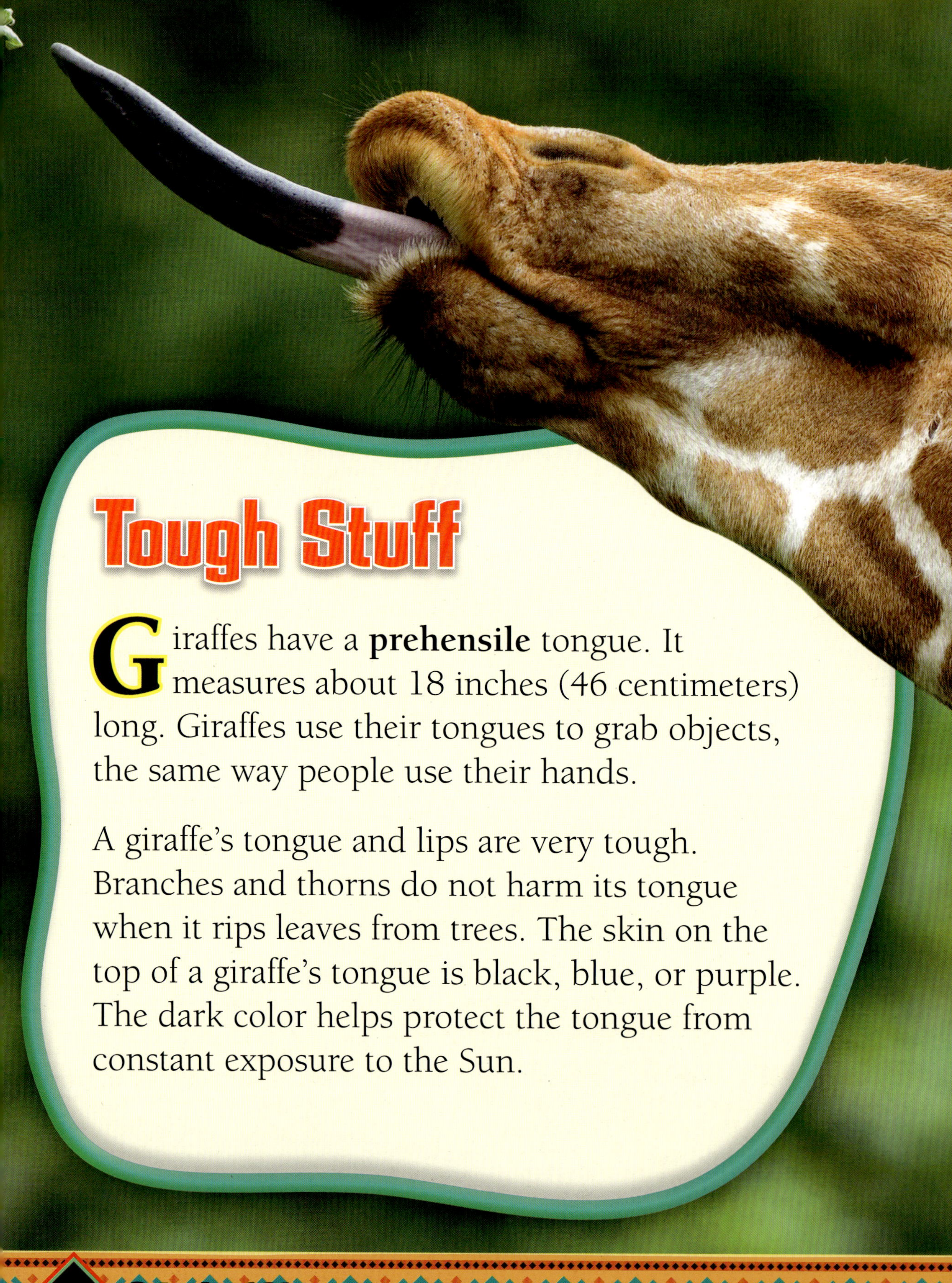

Tough Stuff

Giraffes have a **prehensile** tongue. It measures about 18 inches (46 centimeters) long. Giraffes use their tongues to grab objects, the same way people use their hands.

A giraffe's tongue and lips are very tough. Branches and thorns do not harm its tongue when it rips leaves from trees. The skin on the top of a giraffe's tongue is black, blue, or purple. The dark color helps protect the tongue from constant exposure to the Sun.

BIG FACT

A giraffe's tongue is about **twice as long** as a cat's tail.

The Long and Short

A giraffe's front legs are longer than its back legs. The difference in height between the front and back legs is about 10 percent. This gives the giraffe's back a sloping appearance.

Giraffes use their strong legs to defend themselves by kicking at predators, such as lions. Although they sometimes sleep with their feet tucked underneath their bodies, giraffes also sleep standing up. This helps them stay prepared for danger.

A giraffe's kick is strong enough to kill a lion.

Meal Time

Giraffes are herbivores. This means that they only eat plants. Although giraffes mainly eat leaves, they will also eat other parts of plants that are easy to reach. These include buds, vines, flowers, and fruit.

Plants are a giraffe's main source of water. A giraffe can go for weeks without drinking any water at all. When it does drink, it must put its head down and spread its legs. This position makes the giraffe vulnerable to predator attacks.

If a group of giraffes is at a water hole, some giraffes will stay upright while others drink.

Giraffes in Danger

Throughout history, people have hunted giraffes for their coats. Overhunting has caused the giraffe population to decline. Giraffes are also losing habitat to **development**. Overhunting and loss of habitat have become the biggest threats to giraffes.

There are only about 111,000 giraffes in nature, and their population continues to decrease. Giraffes used to live all over the African continent. Now, they are mainly located in southern and eastern Africa.

People from all over the world travel to Africa to see giraffes in their natural habitat.

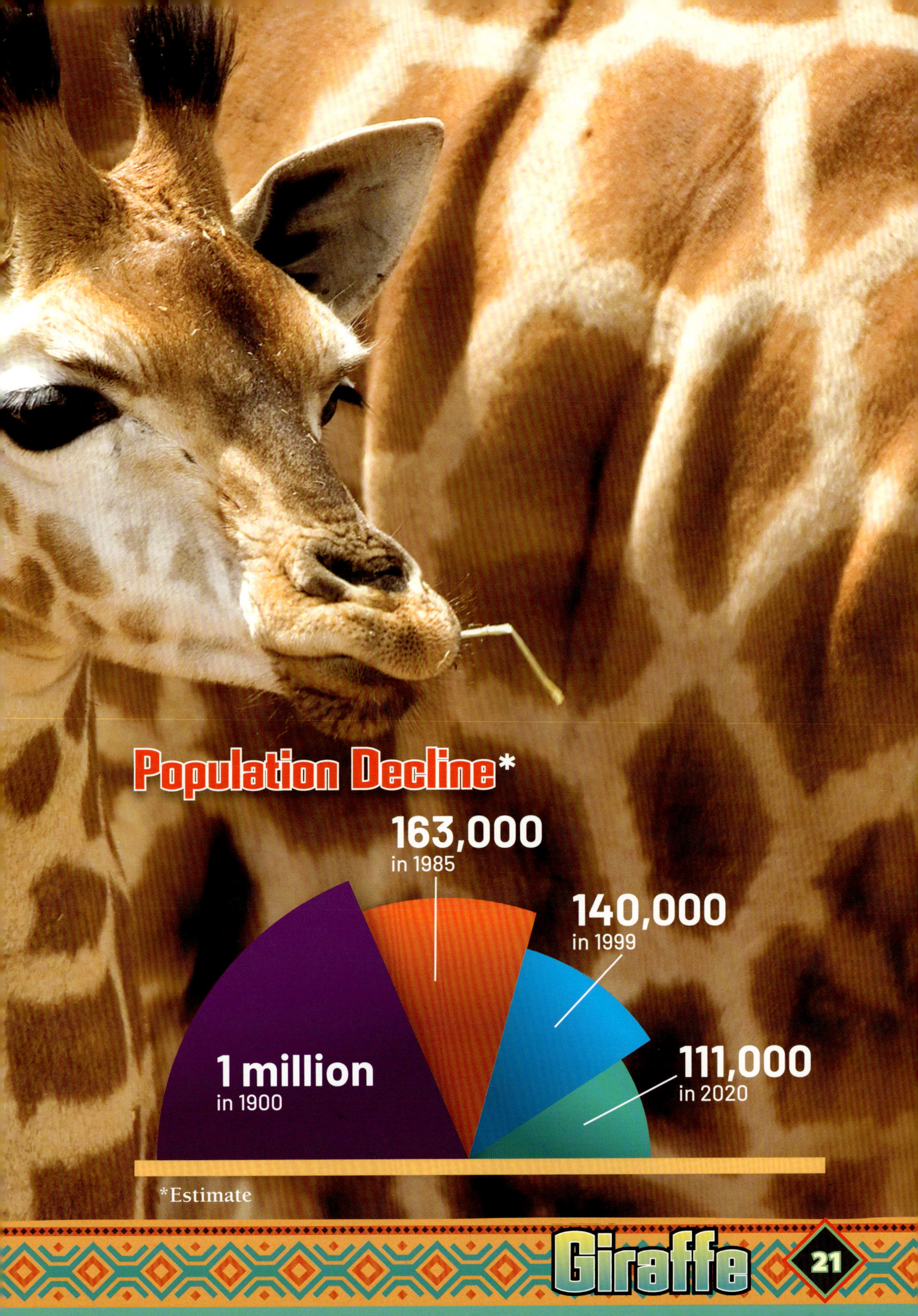
Population Decline*
163,000
in 1985
140,000
in 1999
1 million
in 1900
111,000
in 2020
*Estimate

Giraffe Quiz

1 How many species of giraffe are there?

2 How tall can a giraffe grow to be?

3 How fast can a giraffe run?

4 What are baby giraffes called?

5 What are the bony growths on a giraffe's head called?

6 How long is a giraffe's tongue?

7 Where do giraffes get most of their water?

8 What are the two biggest threats to giraffes?

ANSWERS 1. One **2.** 18 feet (5.5 m) **3.** Up to 35 miles (56 km) per hour **4.** Calves **5.** Ossicones **6.** About 18 inches (46 cm) long **7.** From plants **8.** Overhunting and loss of habitat

Key Words

adapted: changed to suit the environment

development: the placement of structures on land for human use

habitat: the place where an animal lives, grows, and raises its young

predators: animals that live by killing other animals for food

prehensile: capable of grasping

solitary: living or being alone

species: a group of organisms that share features

subspecies: a group within a species that takes on some new characteristics

Index

Get the best of both worlds.

AV2 bridges the gap between print and digital.

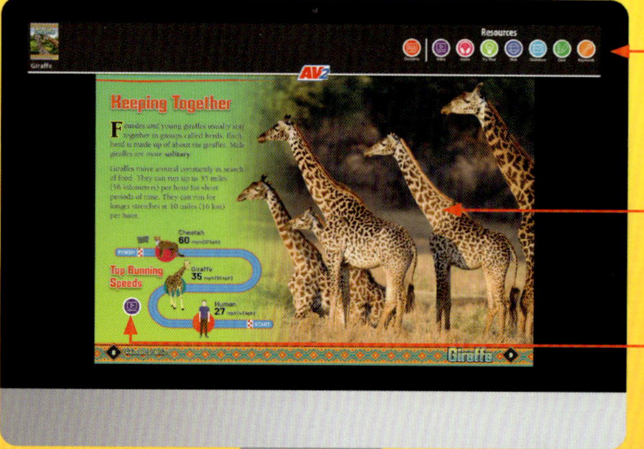

The expandable resources toolbar enables quick access to content including **videos**, **audio**, **activities**, **weblinks**, **slideshows**, **quizzes**, and **key words**.

Animated videos make static images come alive.

Resource icons on each page help readers to further **explore key concepts**.

Published by AV2
276 5th Avenue, Suite 704 #917
New York, NY 10001
Website: www.av2books.com

Library of Congress Cataloging-in-Publication Data
Names: Gillespie, Katie, author.
Title: Giraffe / Katie Gillespie.
Description: New York, NY : AV2, [2022] | Series: Animals of Africa | Includes index. | Audience: Ages 8-11 | Audience: Grades 2-3 | Summary: "Learn about some of the world's most fascinating animals in Animals of Africa"-- Provided by publisher.
Identifiers: LCCN 2020053944 (print) | LCCN 2020053945 (ebook) | ISBN 9781791135188 (library binding) | ISBN 9781791135195 (paperback) | ISBN 9781791135201
Subjects: LCSH: Giraffe--Juvenile literature.
Classification: LCC QL737.U56 G55 2022 (print) | LCC QL737.U56 (ebook) | DDC 599.638--dc23
LC record available at https://lccn.loc.gov/2020053944
LC ebook record available at https://lccn.loc.gov/2020053945

Printed in Guangzhou, China
1 2 3 4 5 6 7 8 9 0 25 24 23 22 21

012021
101120

Project Coordinator: Heather Kissock
Designer: Terry Paulhus

Photo Credits
Every reasonable effort has been made to trace ownership and to obtain permission to reprint copyright material. The publisher would be pleased to have any errors or omissions brought to its attention so that they may be corrected in subsequent printings. AV2 acknowledges Getty Images and Alamy as its primary image suppliers for this title.